MIX
Papier aus verantwortungsvollen Quellen
Paper from responsible sources
FSC® C105338

Shohreh Monshizadeh
Azita Monshizadeh

Computer Application in Electronic Engineering

MATLAB

Anchor Academic
Publishing

Monshizadeh, Shohreh, Monshizadeh, Azita: Computer Application in Electronic
Engineering. MATLAB, Hamburg, Anchor Academic Publishing 2016

Buch-ISBN: 978-3-96067-000-1
PDF-eBook-ISBN: 978-3-96067-500-6
Druck/Herstellung: Anchor Academic Publishing, Hamburg, 2016

Bibliografische Information der Deutschen Nationalbibliothek:
Die Deutsche Nationalbibliothek verzeichnet diese Publikation in der Deutschen
Nationalbibliografie; detaillierte bibliografische Daten sind im Internet über
http://dnb.d-nb.de abrufbar.

Bibliographical Information of the German National Library:
The German National Library lists this publication in the German National Bibliography.
Detailed bibliographic data can be found at: http://dnb.d-nb.de

All rights reserved. This publication may not be reproduced, stored in a retrieval system
or transmitted, in any form or by any means, electronic, mechanical, photocopying,
recording or otherwise, without the prior permission of the publishers.

Das Werk einschließlich aller seiner Teile ist urheberrechtlich geschützt. Jede Verwertung
außerhalb der Grenzen des Urheberrechtsgesetzes ist ohne Zustimmung des Verlages
unzulässig und strafbar. Dies gilt insbesondere für Vervielfältigungen, Übersetzungen,
Mikroverfilmungen und die Einspeicherung und Bearbeitung in elektronischen Systemen.

Die Wiedergabe von Gebrauchsnamen, Handelsnamen, Warenbezeichnungen usw. in
diesem Werk berechtigt auch ohne besondere Kennzeichnung nicht zu der Annahme,
dass solche Namen im Sinne der Warenzeichen- und Markenschutz-Gesetzgebung als frei
zu betrachten wären und daher von jedermann benutzt werden dürften.

Die Informationen in diesem Werk wurden mit Sorgfalt erarbeitet. Dennoch können
Fehler nicht vollständig ausgeschlossen werden und die Diplomica Verlag GmbH, die
Autoren oder Übersetzer übernehmen keine juristische Verantwortung oder irgendeine
Haftung für evtl. verbliebene fehlerhafte Angaben und deren Folgen.

Alle Rechte vorbehalten

© Anchor Academic Publishing, Imprint der Diplomica Verlag GmbH
Hermannstal 119k, 22119 Hamburg
http://www.diplomica-verlag.de, Hamburg 2016
Printed in Germany

Acknowledgments

Dedicated to my Mom
for
her unfailing love and encouragement,
who encouraged me to finish the book in record time. With equanimity and understanding, she stood by me during the endless hours I spent writing.

PREFACE

MATLAB is a numeric computation software for engineering and scientific calculations. MATLAB is increasingly being used by students, researchers, practicing engineers and technicians. The causes of MATLAB popularity are legion. Among them are its iterative mode of operation, built-in functions, simple programming, rich set of graphing facilities, possibilities for writing additional functions, and its extensive toolboxes.

This book explains everything you need to know to begin using MATLAB to do all these things and more. Intermediate and advanced users will find useful information here, especially if they are making the switch to MATLAB 7 from an earlier version.

The book is divided into five parts: Introduction to MATLAB, Calculation and graphs , Programming in MATLAB, Simulation with MATLAB, Circuit analysis applications using MATLAB. It is recommended that the reader work through and experiment with the examples at a computer while reading Chapters 2, 3, and 5. The hands-on approach is one of the best ways of learning MATLAB.

Shohreh Monshizadeh was born in Kermanshah, Iran. She received her Master's education in Power Electronic Engineering at the Islamic Azad University-South Tehran Branch. She had studied in various areas including VFTO Studies Due To The Switching Operation in GIS 132kv Substation And Effective Factors In Reducing These Over Voltages, Reliability Assessment of Power Generation Systems in Presence of Wind Farms Using Fuzzy Logic Method, evaluation reliability in wind farms and Harmonic, unbalance load and reactive power compensation using a new design of passive power factor. She published three journal articles and one book in electronic engineering. Shohreh Monshizadeh is a Professor at Department of Electrical Engineering, Islamic Azad University, Sama College, Kermanshah, Iran.

Azita Monshizadeh was born in Kermanshah, Iran. She received her education in Mathematics at Mathematics Department, Islamic Azad University, (IAU), Kermanshah, Iran

Table of contents

Chapter 1

1 Introduction 6

    1.1 Introduction 7

    1.2 Familiar with MATLAB software environment 7

    1.3 MATLAB Help..10

    1.4 Demo...10

    1.5 Organization of Book..10

Chapter 2

2 Calculation and graphs..12

    2.1 Introduction 13

    2.2 Using MATLAB as a calculator 13

    2.3 Error messagaes...14

    2.4 Miscellaneous commands 15

        2.4.1 Naming variables 15

    2.5 Overwriting variable..16

    2.6 The Colon Operator 17

    2.7 Simple mathematical calculations 17

    2.8 Matrices..19

    2.9 Typing Matrices...20

    2.10 Extracting a Sub-Matrix..21

    2.11 Matrix arithmetic operations...22

    2.12 Array arithmetic operations..23

    2.13 Matrix functions...25

    2.14 Solving linear equations..27

    2.15 Vector drawing (command plot) 28

    2.16 Specifying line styles and colors...31

    2.17 Three-Dimensional Plots...34

Chapter 3
3 Programming in MATLAB..37
    3.1 Programming in MATLAB...38
    3.2 Logical Expressions..38
    3.3 Operator precedence..39
    3.4 Control flow..39
        3.4.1 The "if" statements...39
        3.4.2 The "Switch" statement...42
    3.5 Loops..43
        3.5.1 The "For...end" loop...43
        3.5.2 The "while...end" loop..45

Chapter 4
4 Simulink in MATLAB..47
    4.1 What is simulink...48
    4.2 Simulation..49

Chapter 5
5 Circuit Analysis using MATLAB...63
    5.1 DC Analysis..64
        5.1.1 Node Analysis...64
        5.1.2 Loop Analysis...66

References 74

CHAPTER 1

Introduction

1.1 Introduction

The name MATLAB stands for MATrix LABoratory. MATLAB was written originally to provide easy access to matrix software developed by the LINPACK (linear system package) and EISPACK (Eigen system package) projects[3].

Matlab is a software package that lets you do mathematics and computation, analyse data, develop algorithms, do simulation and modelling, and produce graphical displays and graphical user interfaces[1].

MATLAB is a high-performance language for technical computing. It integrates computation,visualization, and programming environment. Furthermore, MATLAB is a modern programming language environment: it has sophisticated data structures, contains built-in editing and debugging tools , and supports object-oriented programming. These factors make MATLAB an excellent tool for teaching and research[3].

MATLAB has many advantages compared to conventional computer languages (e.g., C, FORTRAN) for solving technical problems. MATLAB is an interactive system whose basic data element is an array that does not require dimensioning. The software package has been commercially available since 1984 and is now considered as a standard tool at most universities and industries worldwide[3].

It has powerful built in routines that enable a very wide variety of computations. It also has easy to use graphics commands that make the visualization of results immediately available. Specific applications are collected in packages referred to as toolbox. There are toolboxes for signal processing, symbolic computation, control theory, simulation, optimization, and several other fields of applied science and engineering[3].

1.2 Familiar with MATLAB software environment

To run matlab on a PC double-click on the matlab icon. To run matlab on a unix system, type matlab at the prompt.

You get matlab to do things for you by typing in commands. matlab prompts you with two greater-than signs (>>) when it is ready to accept a Command from you.

To end a matlab session type quit or exit at the matlab prompt. You can type help at the matlab prompt, or pull down the Help menu on a PC[3].

When starting matlab you should see a following window:

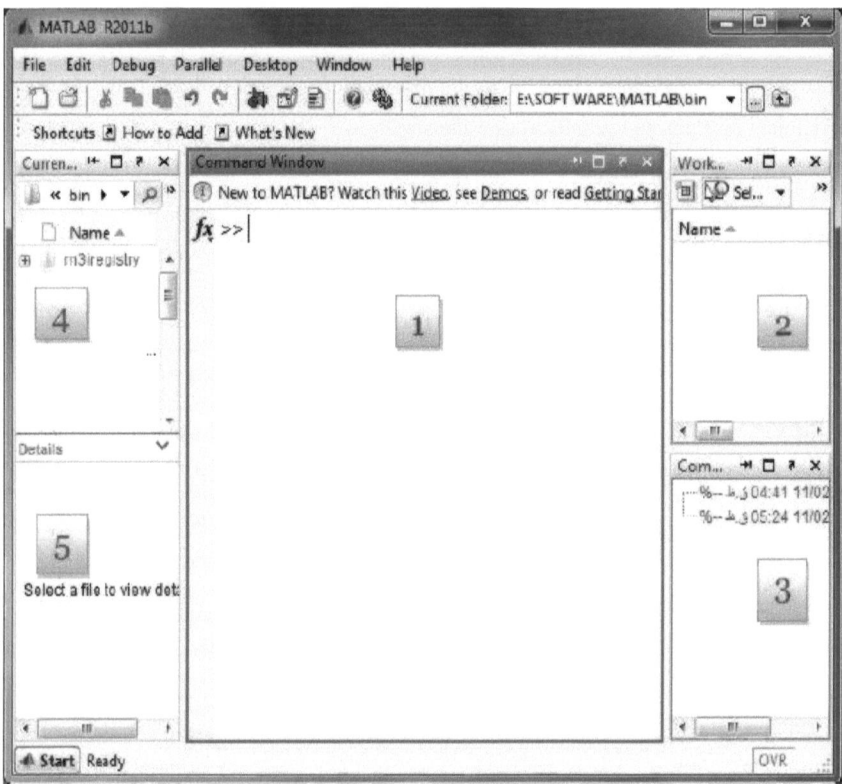

Fig. 1.1: The graphical interface to the MATLAB workspace

As you can see, there are five main sub-windows in this window:

1) Commond window

The window where you type commands and non-graphic output is displayed. A '>>' prompt shows you the system is ready for input. The lower left hand corner of the main window also displays `Ready' or `Busy' when the system is waiting or calculating. Previous commands can be accessed using the up arrow to save typing and reduce errors. Typing a few characters restricts this function to commands beginning with those characters[4].

2) Workspace

Shows the all the variables that you have currently defined and some basic information about each one, including its dimensions, minimum, and maximum values. The icons at the top of the window allow you to perform various basic tasks on variables, creating, saving, deleting, plotting, etc. Double-clicking on a variable opens it in the Variable or Array Editor. All the variables that you've defined can be saved from one session to another using File > Save Workspace As (Ctrl-S). The extension for a workspace file is .mat[4].

3) Command History

Records commands given that session and recent sessions. Can be used for reference or to copy and paste commands[4].

4) Current directory

The directory (folder) that MATLAB is currently working in. This is where anything you save will go by default, and it will also inuence what files MATLAB can see[4].

You won't be able to run a script that you saved that you saved in a diferent directory (unless you give the full directory path), but you can run one that's in a sub-directory. The Current Directory bar at the top centre of the main window lets you change directory in the usual fashion | you can also use the UNIX commands cd and pwd to navigate through directories. The Current Directory window shows a list of all the files in the current directory[4].

5) Details

Details of the selected program will show on the current directory that it is include the type of program, the language of programming and the huge of program[4].

1.3 MATLAB Help

MATLAB's help documentation is very good, and can tell you pretty much anything you need to know. Help>Product Help opens the Help Window, which works largely like a web browser, including forward and back buttons. Use the Contents tab for help oriented around a broad topic (most of what you need will be under the MATLAB heading, and then probably Getting Starting or Graphics) Search or Index for more specific queries (e.g. inter- polating values, polynomial fit, etc.). The `see also' at the end of each file is very useful if you haven't found quite the right thing. It can also suggest better ways of doing things. Typing help commandname in the Command Window will also bring up the help file for that command[4].

1.4 Demo

The matlab demos are well worth browsing. You can learn about a subject (often reading references are given), as well as learning about matlab's capabilities. Of interest to sonar and radar signal processors is matlab's Higher Order Spectral Analysis toolbox containing, for example, functions for direction of arrival estimation (beamforming plus other methods), time-frequency distributions, and harmonic estimation. Type help hosa for a list of functions in the Higher Order Spectral Analysis toolbox. Browsing the demos or doing a keyword search may save you from writing your own matlab code and re-inventing the wheel[5].

If you like to see a demo of MATLAB you can type demo after the prompt.

1.5 Organization of Book

In writing, we drew on our experience to provide important information as quickly as possible. The book contains a short, focused introduction to MATLAB and application of computer in Electronic engineering. It contains practice problems (withcomplete solutions) so you can test your knowledge. There are several illuminating sample projects that show you how MATLAB can be used in real-world applications, and there is an entire chapter on troubleshooting.

Here is a detailed summary of the contents of the book:

Chapter 1, Getting Started, describes how to start MATLAB. It tells you how to enter commands, how to access online help, how to recognize the various MATLAB windows you will encounter, and how to exit the application.

Chapter 2, MATLAB basics, shows you how to do elementary mathematics using MATLAB. This chapter contains the most essential MATLAB commands.

Chapter 3, Interacting with MATLAB, contains an introduction to the MATLAB Desktop interface. This chapter will introduce you to the basic window features of the application, to the small program files (M-files) that you will use to make most effective use of the software, and to a simple method (diary files) of documenting your MATLAB sessions. After completing this chapter, you'll have a better appreciation of the breadth described in the quote that opens this preface.

Chapter 4, SIMULINK and GUIs, consists of two parts. The first part describes the MATLAB companion software SIMULINK, a graphically oriented package for modeling, simulating, and analyzing dynamical systems. Many of the calculations that can be done with MATLAB can be done equally well with SIMULINK.

Chapter 5, In this chapter described the Circuit Analysis Using Matlab that consists of DC analysis. In DC analysis of circuits both node analysis and loop analysis of circuit are considered.

CHAPTER 2

Calculations and graphs

2.1 Introduction

Matlab (Matrix laboratory) is an interactive software system for numerical computations and graphics. As the name suggests, Matlab is especially designed for matrix computations: solving systems of linear equations, computing eigenvalues and eigenvectors, factoring matrices, and so forth. In addition, it has a variety of graphical capabilities, and can be extended through programs written in its own programming language. Many such programs come with the system; a number of these extend Matlab's capabilities to nonlinear problems, such as the solution of initial value problems for ordinary differential equations[1].

Matlab is designed to solve problems numerically, that is, in finite-precision arithmetic. Therefore it produces approximate rather than exact solutions, and should not be confused with a symbolic computation system (SCS) such as Mathematica or Maple. It should be understood that this does not make Matlab better or worse than an SCS; it is a tool designed for different tasks and is therefore not directly comparable[2],[6].

In the following sections, I give an introduction to some of the most useful features of Matlab. I include plenty of examples; the best way to learn to use Matlab is to read this in front of a computer, trying the examples and experimenting.

2.2 Using MATLAB as a calculator

You are now faced with the MATLAB desktop on your computer, which contains the prompt (>>) in the Command Window. Usually, there are 2 types of prompt[3]:

>> for full version

EDU> for educational version

Note: To simplify the notation, we will use this prompt, >>, as a standard prompt sign, though our MATLAB version is for educational purpose[3].

You will have noticed that if you do not specify an output variable, MATLAB uses a default variable ans, short for answer, to store the results of the current calculation[3]. Note that the variable ans is created (or overwritten, if it is already existed). To avoid this, you may assign a value to a variable or output argument name. For example,

>> x = 1+5*3

ans=

16

will result in x being given the value 1 + 5*3 = 16. This variable name can always be used to refer to the results of the previous computations. Therefore, computing 4x will result in

>> 4*x

ans =

64.0000

For mathematics calculation in MATLAB, there are some operators that Table 2.1 gives the partial list of arithmetic operators[3].

Table 2.1: Basic arithmetic operators

Symbol	Operation	Example
+	Addition	5+3
-	Subtraction	5-3
*	Multiplication	5*3
/	Division	5/3

2.3 Error messages

If we enter an expression incorrectly, MATLAB will return an error message[3]. For example, in the following, we left out the multiplication sign, *, in the following expression:

>> x = 10;

>> 8x

??? 8x

Error: Unexpected MATLAB expression.

This is mean that 8x is unfamiliar with MATLAB software.

2.4 Miscellaneous commands

Here are few additional useful commands for clearing command window as following[3]:

clc: Clear the Command Window

clear x: Remove x from the workspace

clear (all): Removes all variables from the workspace

2.4.1 Naming variables

At naming variables must respected the following items:
- MATLAB is sensitive in case of toward upper and lower letters
- Variable names are up to 31 characters
- Variable names must be begin with letter, not number.
- they must not be one of the keywords of MATLAB, for showing the keywords, the iskeywords command must be used. If function is the keyword, the ans of this function is 1 and else it is not the keywords the ans of fuction is 0,

for example

\>\> iskeyword('for')

ans= 1

\>\> iskeyword('number')

ans= 0

MATLAB can solve the simplify mathematics operators such as one calculator. It is easy use MATLAB as calculator. The symbols +,- and / have their usual meaning, * denotes multiplication and ^ exponentiation[6].

\>\> 1+2*3

ans =

7

Or it can be the values saved in variables and calculations are performed using letters as following:

\>> a= 5

a=5

\>> b=8

b=8

\>> a+b

ans=
   8

2.5 Overwriting variable

Once a variable has been created, it can be reassigned. In addition, if you do not wish to see the intermediate results, you can suppress the numerical output by putting a semicolon (;) at the end of the line[5]. Then the sequence of commands looks like this:

\>> t = 5;

\>> t = t+1

t =

6

It is possible to enter multiple statements per line. Use commas (,) or semicolons (;) to enter more than one statement at once. Commas (,) allow multiple statements per line without suppressing output[3].

\>> a=7; b=cos(a), c=cosh(a)

b =

0.6570

c =

548.3170

2.6 The Colon Operator

To generate a vector of equally-spaced elements matlab provides the colon operator. Try the following commands[7]:

1:5

0:2:10

0:.1:2*pi

The syntax *x:y* means roughly "generate the ordered set of numbers from *x* to *y* with increment 1 between them." The syntax *x:d:y* means roughly "generate the ordered set of numbers from *x* to *y* with increment *d* between them"[7].

2.7 Simple mathematical calculations

For the calculation of simple mathematical calculations,expressions are written in the command window and press the Enter key. As mentioned above, calculation are done both numerical and variables that we use the second method for storing values in the variables. There ae many Predefined variables and math constants in the MATLAB that each of them have espacial meaning as folowing[3]:

Table 2.2: Special Variables and Constants

Phrase	meaning
ans	Value of last variable (answer)
pi	The number π (3/14159)
eps	Floating-point relative accuracy
exp	Natural logarithm
inf	Infinity(∞)
NAN	Not a number
syms	Declaring the variables
Expand	multiply out the expression
factor	forced MATLAB to restore it to factored form
simplify	you can sometimes use to express a formula as simply as possible
solve	Solve the problems
sqrt	Sqrt from one number
sqrtm	Sqrtm from one Matrix
numel	Number of elements in a matrix
i	Imaginary unit of a complex number
Realmin	The smallest positive integer
realmax	The largest positive integer
factorial	Canculate the factorial one number

There is one example of above Predefined variables and math constants as folowing:

Here we have three variables with different values:

\>\> a=(1/2)*pi;

\>\> b=(5/4)*pi;

\>\> c=pi;

\>\> eps(a)

ans=

2.2204e-016

\>\> exp(b)

ans=

10.5507

Using MATLAB's Symbolic MathT oolbox, you can carry out algebraic or symbolic calculations suchas factoring polynomials or solving algebraic equations. Type help symbolic to make sure that the Symbolic Math Toolbox is installed on your system.

To perform symbolic computations, you must use syms to declare the variables you plan to use to be symbolic variables. Consider the following series of commands:

\>\> syms x y

\>\> (x - y)*(x - y)^2

ans =

(x-y)^3

Using the various mathematical operations with variables can be done as following:

\>\> expand((a^2+b^3)-(a^3+c^2))

ans=

-a^3+a^2+b^3-c^2

```
>> factor((a^2+b^3)-(a^3+c^2))
ans=

-(a^3 - a^2 - b^3 + c^2)

>> simplify((a^2+b^3)-(a^3+c^2))
ans=

- a^3 + a^2 + b^3 - c^2

>>solve((a^2+b^3)-(a^3+c^2))
ans=

 (- a^3 + a^2 + b^3)^(1/2)
-(- a^3 + a^2 + b^3)^(1/2)

>> realmin
ans=

2.225073858507201e-308

>> realmax
ans=

1.797693134862316e+308

>> factorial(5)
ans=

120
```

2.8 Matrices

The basic object that matlab deals with is a matrix. A matrix is an array of numbers. For example the following are matrices[3]:

- A matrix with only one row is called a row vector. A row vector can be created in Matlab as follows (note the commas)[3]:

\>> rowvec = [12 , 14 , 63]

rowvec =

12 14 63

- A matrix with only one column is called a column vector. A column vector can be created in MATLAB as follows (note the semicolons)[3]:

colvec = [13 ; 45 ; -2]
colvec =
13
45
-2

- A matrix can be created in Matlab as follows (Notice that the rows of a matrix are separated by semicolons, while the entries on a row are separated by spaces or commas)[3]:

matrix = [1 , 2 , 3 ; 4 , 5 ,6 ; 7 , 8 , 9]
matrix =

1 2 3
4 5 6
7 8 9

2.9 Typing Matrices

To type a matrix into matlab you must[4]:
- begin with a square bracket [
- separate elements in a row with commas or spaces
- use a semicolon ; to separate rows
- end the matrix with another square bracket].

For example type:

a = [1 2 3;4 5 6;7 8 9]

matlab responds with

a =

1 2 3
4 5 6
7 8 9

2.10 Extracting a Sub-Matrix

- A portion of a matrix can be extracted and stored in a smaller matrix by specifying the names of both matrices and the rows and columns to extract[3]. The syntax is:

    sub_matrix = matrix (r1 : r2 , c1 : c2) ;

where r1 and r2 specify the beginning and ending rows and c1 and c2 specify the beginning and ending columns to be extracted to make the new matrix[3].
for example:
A column vector can be extracted from a matrix. As an example we create a matrix below:

>> matrix=[1,2,3;4,5,6;7,8,9]

matrix =

1 2 3
4 5 6
7 8 9

Here we extract column 2 of the matrix and make a column vector:

>> col_two=matrix(: , 2)

col_two =

2
5
8

A row vector can be extracted from a matrix. As an example we create a matrix below:

» matrix=[1,2,3;4,5,6;7,8,9]

matrix =

   1 2 3
   4 5 6
   7 8 9

Here we extract row 2 of the matrix and make a row vector. Note that the 2:2 specifies the second row and the 1:3 specifies which columns of the row.

>> rowvec=matrix(2 : 2 , 1 : 3)

rowvec =

   4 5 6

2.11 Matrix arithmetic operations

As we mentioned earlier, MATLAB allows arithmetic operations: + , - , * , and ^ to be carried out on matrices[6]. Thus,

A+B or B+A is valid if A and B are of the same size

A*B is valid if A's number of column equals B's number of rows

A^2 is valid if A is square and equals A*A

α*A or A*α multiplies each element of A by α

A/B : division A on B

A-b: substract A of B

You can multiply two matrices together using the * operator:

>> a = [1 2 3;2 3 4; 3 4 1]

a =

   1 2 3
   2 3 4
   3 4 1

>> b = [1 0 7;2 -3 4;3 -4 3]

b =

1 0 7

2 -3 4

3 -4 3

>> a*b

ans =

 14 -18 24

 20 -25 38

 14 -16 40

>> a/b

ans=

        1.571428571428571   -3.714285714285714 2.285714285714285

        2.571428571428571   -6.714285714285713 4.285714285714284

        3.285714285714285   -10.857142857142854 7.142857142857141

>>a+b

ans=

 2 2 10

 4 0 8

 6 0 4

>> a-b

ans=

 0 2 -4

 0 6 0

 0 8 -2

2.12 Array arithmetic operations

On the other hand, array arithmetic operations or array operations for short, are done element-by-element. The period character, ., distinguishes the array operations from the matrix operations. However, since the matrix and array operations are the same for addition (+) and subtraction (−), the character pairs (*) and (/) are not used[3]. The list of array operators is shown below in Table 2.3. Summary of matrix and array operations as folowing:

Table 2.3: Summary of matrix and array operations[3]

Operation	MATRIX	ARRAY
Addition	+	+
Subtraction	-	-
Multiplication	*	.*
Division	/	./
Left division	\	\.
Exponentiation	^	.^

If A and B are two matrices of the same size with elements A = [a_{ij}] and B = [b_{ij}], then the command[4].

* Element-by-element multiplication

./ Element-by-element division

.^ Element-by-element exponentiation

To multiply the elements of two matrices use the .* operator:

\>\> a = [1 2;3 4]

a =

  1 2

  3 4

\>\> b = [2 3;0 1]

b =

  2 3

  0 1

\>\> a.*b

ans =

```
2  6
0  4
```

\>\> a./b

ans=

```
0.5000000000  0.66666666666
     inf       4.00000000000
```

2.13 Matrix functions

MATLAB provides many matrix functions for various matrix/vector manipulations; see Table 2.4 for some of these functions. Use the online help of MATLAB to find how to use these functions[4].

Table 2.4: Matrix functions

function	operation
det	Determinant
inv	Matrix inverse
pinv	Non-square matrix inverse
trace	Sum of numbers from diagonal matrix
max	The largest element of each column of matrix
min	The smallest element of each column of Matrix
sum	Collected rows and columns of the matrix
prod	Multiplying element of Matrix
numel	The number of elements of Matrix
Ones( )	Create array of all ones

There are some functions such as above operations as folowing:

\>\> a=[1 2 3;2 3 4;3 4 1];

\>\> det(a)

ans=

   4

\>\>inv(a)

ans=

-3.25 2.50 -0.25

2.50 -2.00 0.50
-0.25 0.50 -0.25

\>> pinv(a)

ans=

-3.25 2.50 -0.25
2.50 -2.00 0.50
-0.25 0.50 -0.25

\>> trace(a)

ans=

　5

\>> max(a)

ans=

　3 1 1

\>> min(a)

ans=

1 2 1

\>> sum(a)

ans=

6 9 8

\>> prod(a)

ans=

6 24 12

\>> numel(a)

ans=

　9

```
>>ones(3)
ans=
   1 1 1
   1 1 1
   1 1 1
```

2.14 Solving linear equations

One of the problems encountered most frequently in scientific computation is the solution of systems of simultaneous linear equations. With matrix notation, a system of simultaneous linear equations is written[3]:

$$Ax=b$$

where there are as many equations as unknown. A is a given square matrix of order n, b is a given column vector of n components, and x is an unknown column vector of n components[3].

In linear algebra we learn that the solution to $Ax = b$ can be written as $x = A^{-1}b$, where A^{-1} is the inverse of A.

For example, consider the following system of linear equations[3]:

$$\begin{cases} x + 2y = 1 \\ 4x + 5y = 1 \end{cases}$$

The coeficient matrix A is:

$$A = \begin{bmatrix} 1 & 2 \\ 4 & 5 \end{bmatrix} \text{ and the vector } b = \begin{bmatrix} 1 \\ 1 \end{bmatrix}$$

With matrix notation, a system of simultaneous linear equations is written:

$$Ax=b$$

This equation can be solved for x using linear algebra. The result is $x = A^{-1}b$.

There are typically two ways to solve for x in MATLAB:

The first one is to use the matrix inverse, inv.

```
A = [1 2 ; 4 5 ];
>> b = [1; 1];
>> x = inv(A)*b
x =
   -1.0000
    1.0000
```

The second one is to use the *backslash* (\)operator. The numerical algorithm behind this operator is computationally efficient. This is a numerically reliable way of solving system of linear equations by using a well-known process of Gaussian elimination[3].

```
>> A = [1 2 ; 4 5];
>> b = [1; 1];
>> x = A\b
x =
   -1.0000
    1.0000
```

This problem is at the heart of many problems in scientific computation. Hence it is important that we know how to solve this type of problem efficiently.

2.15 Vector drawing (command plot)

Matlab provides a variety of functions for displaying data as 2-D or 3-D graphics.

The basic MATLAB graphing procedure, for example in 2D, is to take a vector of x coordinates, $x = (x_1,...., x_N)$, and a vector of y-coordinates, $y = (y_1,...., y_N)$, , locate the points (x_i, y_i), with $i = 1, 2,....., n$ and then join them by straight lines. You need to prepare x and y in an identical array form; namely, x and y are both row arrays or column arrays of the same length[8],[16].

The MATLAB command to plot a graph is plot(x,y). The vectors $x = (1, 2, 3, 4, 5, 6)$ and $y = (3,-1, 2, 4, 5, 1)$ produce the picture shown in Figure 2.1.

Example 1:
\>\> x = [1 2 3 4 5 6];
\>\> y = [3 -1 2 4 5 1];
\>\> plot(x,y)

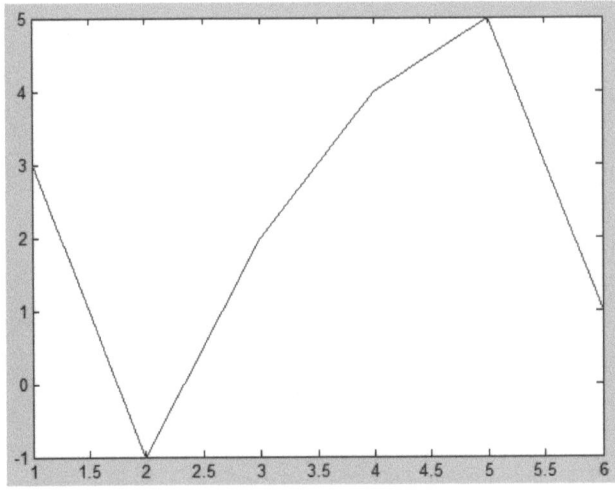

Fig. 2.1: Example 1

For example, to plot the function sin (x) on the interval [0; 2π], we first create a vector of x values ranging from 0 to 2π, then compute the *sine* of these values, and finally plot the result:

Example 2:
\>\> x = 0:pi/100:2*pi;
\>\> y = sin(x);
\>\> plot(x,y)

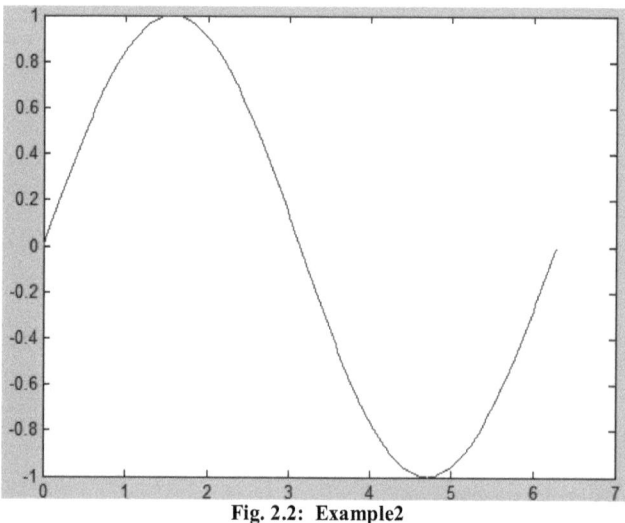

Fig. 2.2: Example2

Notes:

0:pi/100:2*pi yields a vector that

- starts at 0,
- takes steps (or increments) of $\pi/100$,
- stops when 2π is reached.

There are commands in Matlab to "annotate" a plot to put on axis labels, titles, and legends. For example:

\>> % To put a label on the axes we would use:
\>> xlabel ('X-axis label')
\>> ylabel ('Y-axis label')
\>> % To put a title on the plot, we would use:
\>> title ('Title of my plot')

There are several common functions in MATLAB for plotting in the table 2.5, as following[9]:

Table 2.5: Several command Plot function[9]

Function	Action
Hold on	Adding new graph without the remove the previos graph
Hold off	Exitting of hold on
Legend	Legend placement by mouse
Grid	Network command of figure
plot	Generates xy plot
Plot3	3-D plot
Subplot	Draw multiple graphs side by side
Title	Put the name for figuire
xlabel	Adds text label to x-axis
ylabel	Adds text label to y-axis
mesh	Creates three-dimensional mesh surface plot
meshgrid	Creates rectangular grid
figure	Opens a new figure window
text	Places string in figure

2.16 Specifying line styles and colors

It is possible to specify line styles, colors, and markers (e.g., circles, plus signs, . . .) using the plot command[3]:

plot(x,y,'style_color_marker')

where style_color_marker is a triplet of values from Table 2.6. To find additional information, type help plot or doc plot[9].

Table 2.6: Colors, Symbols and Line Types[9]

Color		Symbol		Line	
Blue	b	+	Plus	-	solid
Green	g	0	Circle	:	dotted
Yellow	y	*	Asterisk	-.	Dash-dot
Red	r	.	Piont	- -	Dashed
Black	k	x	x-mark		
Cyan	c	d	diamond		
Magenta	m	^	triangle (up)		
White	w	p	pentagram		

Example 3: We want to draw two plots, in which one is y=sin(x) with blue color and the other is y=cos(x) with red color, the x-label is sin and the y-label is cos. Also, the legend must be applied for figures:

\>\> x=0:0.1:2*pi;
\>\>y=sin(x);
\>\>g=cos(x);
\>\>plot(x,y,'b')
\>\>hold on
\>\>plot(x,g,'r')
\>\>x-label(sin)
\>\>y-label(cos)
\>\>title(figure)
\>\>legend('sinx','cosx')
\>\>text(3.2,0,'sin')
\>\>text(1.65,0,'cos')

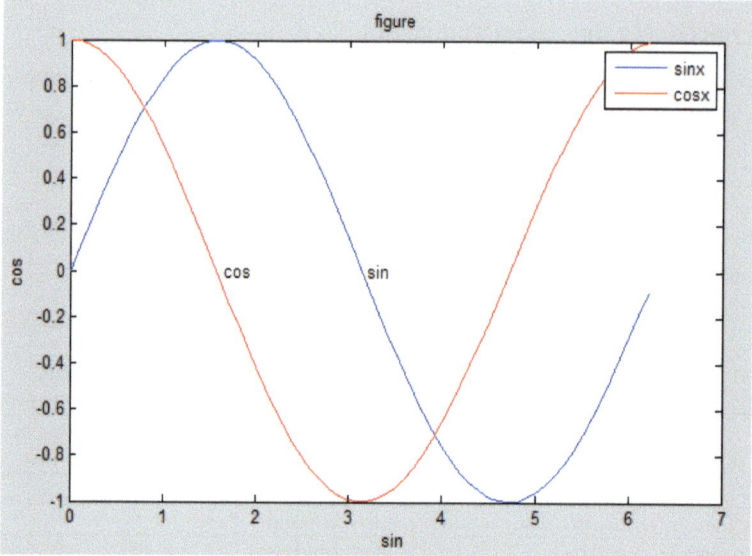

Fig. 2.3: Example 3

The other example, we want to drw two plots that one of them is y=sin(x) with black color and : style and the other is y=cos(x) with red color and -. style:

Example4:

x=-pi:pi/10:pi

y=sin(x);

z=cos(x);

plot(x,y,'b :',x,z,'r -.')

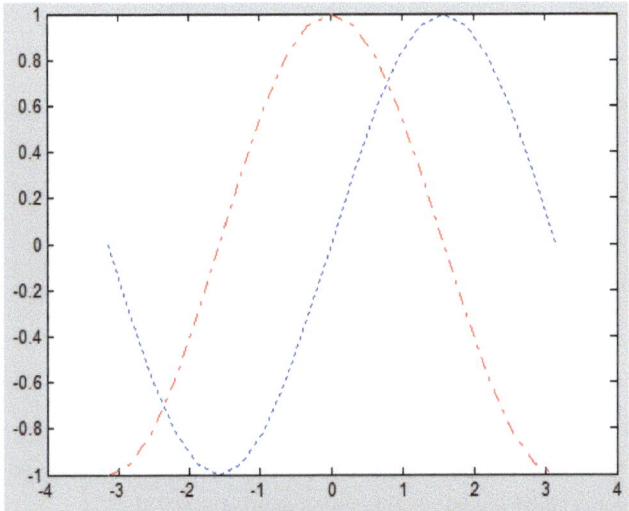

Fig. 2.4: Example 4

Example 5: these statements plot three related functions of x: y1 = 2*cos(x), y2 = cos(x), and y3 =0.5 *cos(x), in the interval $0 \leq x \leq 2\pi$.

\>\> x=0:pi/100:2*pi;

\>\> y1=2*cos(x;(

\>\> y2=cos(x;(

\>\> y3=0.5*cos(x;(

\>\> plot(x,y1,'--',x,y2,'-',x,y3,':')

\>\> xlabel('x')

\>\> ylabel('y')

\>\> title('shape')

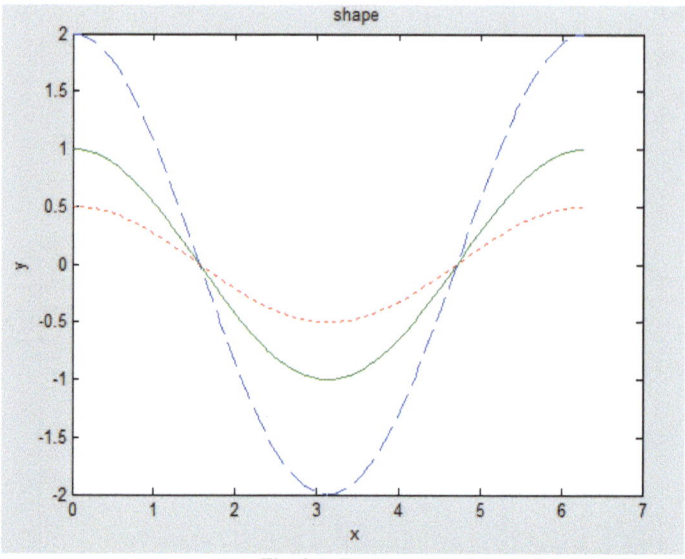

Fig. 2.5: Example 5

2.17 Three-Dimensional Plots

Matlab provides a variety of functions for displaying data as 2-D or 3-D graphics. For 3-D graphics, the most commonly used commands are[8]:

- plot3(x1, y1, z1, 'line style', x2, y2, z2, 'line style'...)
- mesh(x,y,Z), surf(x,y,Z)

The first statement is a three-dimensional analogue of plot( ) and plots lines and points in 3-D[8].

The second statementis for surface or mesh plots, you use the second statement where x, y are vectors or matrices and Z is a matrix.

Example 6: Plot y1 = sin(x) and y2 = cos(x) with x in [-3 π;3π] on the same graph. The first step is to define a set of values for x at which the functions will be defined. Use plot 3-d for showing the above function:

x=-3pi:pi/30:3pi;

x=sin(t)

y=cos(t)

z=t

plot3(x,y,z);

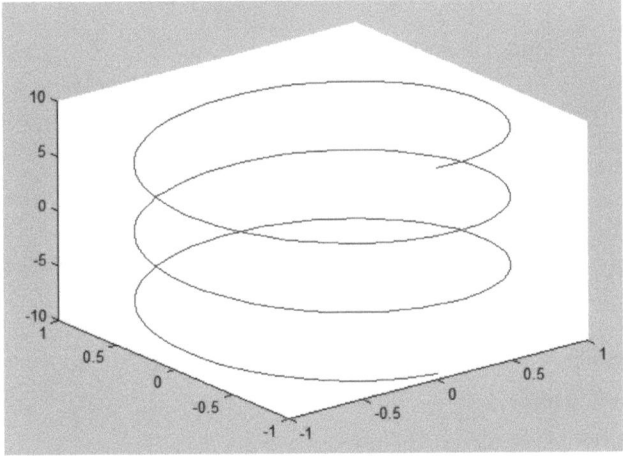

Fig. 2.6: Example 6

Example 7: Plot z = sin(r)=r with r=$\sqrt{x^2 + y^2}$, $-8 \leq x \leq 8, -8 \leq y \leq 8$:

The first step in displaying a function of two variables, z = f(x, y), is to use the meshgrid function to generate X and Y matrices consisting of repeated rows and columns, respectively, over the domain of the function. The function can then be evaluated and graphed[8].

x=-8: 0.5: 8;

y=x;

[x,y]=meshgrid(x,y);

R=sqrt(x.^2+y.^2)+eps;

Z=sin(R) ./R;
mesh(x,y,z)

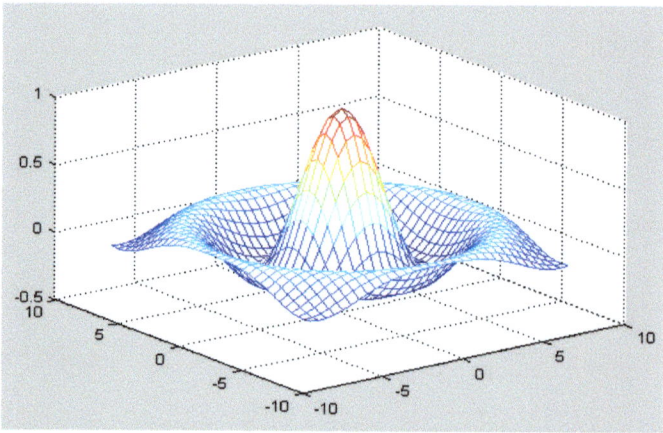

Fig. 2.7: Example 7

Example 8:

x=-8: 0.5: 8;
y=x;
[x,y]=meshgrid(-2:2);
z=sqrt(x.^2+y.^2)+eps;
mesh(x,y,z)

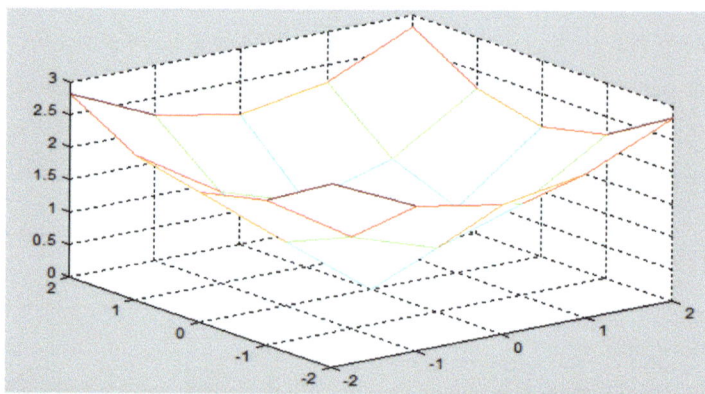

Fig. 2.8: Example 8

CHAPTER 3

Programming in Matlab

3.1 Programming in Matlab

MATLAB is also a programming language. Like other computer programming languages, MATLAB has some decision making structures for control of command execution. These decision making or control fow structures include for loops, while loops, and if-else-end constructions. Control flow structures are often used in script M-files and function M-files[10].

I will first discuss the programming mechanisms and then explain how to write programs.

3.2 Logical Expressions

A relational operator compares two numbers by determining whether a comparison is true or false. Relational operators are shown in Table 3.1[10].

Table 3.1: Relational and logical operators[4]

Operator	Description
>	Greater than
<	Less than
>=	Greater than or equal to
<=	Less than or equal to
==	Equal to
~=	Not Equal to
&	AND operator
| |	OR operator
~	NOT operator

These combine statements which evaluate to either true or false (x <= 10, y == 6, etc.) with the logical operators & (logical AND), | (logical OR), and ~ (logical NOT). Note that when testing for equality, you must use ==. A single = assigns the value of the right hand side to the left hand side. If you are used to other programming languages, && and || also work (and short-circuit, which the single forms do not)[4].

Sample expressions:

x < 10 & y == 4   x is less than 10 and y equals 4 x + y >= 100 | x + y < 50 ~ x == y   x + y is greater than or equal to 100 or less than 50 and x does not equal y

3.3 Operator precedence

We can build expressions that use any combination of arithmetic, relational, and logical operators. Precedence rules determine the order in which MATLAB evaluates an expression. Here we add other operators in the list. The precedence rules for MATLAB are shown in this list (Table 3.2), ordered from highest (1) to lowest (8) precedence level[10].

Table 3.2: Operator precedence[10]

Priority	Operation
1(highest)	Parentheses(If parentheses are involuted, the internal parentheses have higher priority)
2	Power
3	Logical operator (~)
4	Multiplication and division
5	Addition and subtraction
6	Comperasion operatores (=, <, >, <=, >=, ==, ~=)
7	Logical operator AND (&)
8	Logical operator OR(|)

3.4 Control flow

MATLAB has four control flow structures: the if statement, the for loop, the while loop, and the switch statement[9].

3.4.1 The "if" statements

The ``if...end" structure MATLAB supports the variants of "if" construct[9].

- if ... end
- if ... else ... end
- if ... elseif ... else ... end

The simplest form of the if statement is[9]:

if   expression
    statements
end

Here are some examples based on the familiar quadratic formula.

Example 1:

if x < 10

$\quad$ b = 2^x

end

Which says:

If x is less than 10, set b equal to 2^x. If x is greater than or equal to 10, b won't be assigned any value (or will retain the value it had before the if statement started.

Example 2:

Write a programm by using if-end that shows the days of the week by pressing 1 to 7 numbers?

```
clear all
clc
a=input('please enter number of week day:')
if a==1
    display('satarday');
end
if a==2
    display('sunday');
end
if a==3
    display('monday');
end
if a==4
    display('tuesday');
end
if a==5
    display('wednesday');
end
if a==6
    display('tursday');
end
if a==7
    display('friday');
end
```

Example 3: Write a programm by using if-else-end that is determined the zero, negetive and positive of two numbers?

```
clear all
clc
a=input('please inter number a:');
b=input('please inter number b:');
if a>0 && b>0 disp('a and b are positive')
elseif a<0 && b<0 disp('a and b are negetive')
elseif a<0 && b==0 disp('a is negetive and b is zero')
elseif a>0 && b==0 disp('a is positive and b is zero')
elseif a==0 && b>0 disp('a is zero and b is positive')
elseif a==0 && b<0 disp('a is zero and b is negetive')
elseif a>0 && b<0 disp('a is positive and b is negetive')
elseif a<0 && b>0 disp('a is negetive and b is positive')
elseif a==0 && b==0 disp('numbers are zero')
end
```

To evaluate diferent code for each of several possible alternatives, we have[4]:

if (logical expression)

    (code to evaluate if expression is true)

elseif (2nd logical expression)

    (code to evaluate if 2nd expression is true)

...

else

    (code to evaluate if no expression is true)

end

Here, if the first expression is true, the first batch of code is evaluated, and the programme leaves the if block of code. If the first expression is false, it tries the second expression, and so on. Only one batch of code is ever evaluated. You can have as many elsif statements as you need, and you do not have to have an else statement if no code is to be evaluated when all expressions are false[4].

These statements can also be nested - you can put another if statement in the code to be evaluated if a given expression is true,

Example 4:

```
clear all
clc
if x<10
    if y<10
        b=x*y
    elseif
        b=x+y
    else
        b=x
    end
end
```

3.4.2 The "Switch" statement

This case provided another way to control the program:

switch

case 1

    (logical expression)

    (code to evaluate if expression is true)

case 2

    (logical expression)

    (code to evaluate if expression is true)

case 3

….

otherwise

    (logical expression)

    (code to evaluate if expression is true)

end

Example 5:

Write a prograam by using switch case that display the days of week by pressing the numbers from 1 to 7?

```
clear all
clc
a=input('please enter number of week day:')
switch a
    case 1
        disp('satrday')
    case 2
        disp('sunday')
    case 3
        disp('monday')
    case 4
        disp('tuesday')
    case 5
        disp('wedendday')
    case 6
        disp('tursday')
    case 7
        disp('friday')
    otherwise
        disp('your entered is not correct')
end
```

3.5 Loops

Loops are for when you want to execute a statement a set number of times or for as long as some expression is true. The for statement is used to evaluate code a set number of times[4].

3.5.1 The "For...end" loop

In the for ... end loop, the execution of a command is repeated at a fixed and predetermined number of times[4]. The syntax is:

for variable = expression
    statements
end

Usually, expression is a vector of the form i:s:j. A simple example of for loop is

for ii=1:5
    x=ii*ii

end

It is a good idea to indent the loops for readability, especially when they are nested. Note that MATLAB editor does it automatically.

Example 6: x goes from 1 to 20

```
clear all
clc
b=0
for x=1:20
    b=b+x^2
end
```

which Finds the sum of the square from 1 to 20. The b = 0 ensures that any previous value of b doesn't affect the result.

Example 7: By using of For loop, write a program that takes numbers and even numbers print up to 1000.

```
clear all
clc
a=input('inter number:');
for b=a:1:1000
    if mod(b,2)==0
        disp(['ans is:',num2str(b)])
    else
        continue
    end
end
```

Example 8: By using for loop, write a program that takes two numbers and calculate the greatest common divisior of two numbers?

```
clear all
clc
a=input('inter a number:');
b=input('enter other number:')
if a>b
    m=a;
    n=b;
else
    m=b
```

```
        n=a
end
disp(['greater number is:',num2str(m)])
disp(['less number is:',num2str(n)])
for c=n:-1:1
    if mod(m,c)==0 && mod(n,c)==0
        break
    end
end
disp(['B.M.M is:', num2str(c)])
```

3.5.2 The "while...end" loop

This loop is used when the number of passes is not specified. The looping continues until a stated condition is satisfied. The while loop has the form[9]:

while   expression

       statements

end

The statements are executed as long as expression is true[9].

x = 1

while x <= 10

     x = 3*x

end

It is important to note that if the condition inside the looping is not well defined, the looping will continue indefinitely. If this happens, we can stop the execution by pressing Ctrl-C[9].

Example 9:
```
clear all
clc
x=1
while x<2000
    x=2*x
end
```

Which Prints the powers of 2 that are less than 2000. The x = 1 starts the process off correctly.

Example 10: By using while loop, write a program that takes two numbers and calculate the greatest common divisior of two numbers?

```
clear all
clc
a=input('inter a number:');
b=input('enter other number:')
if a>b
    m=a;
    n=b;
else
    m=b
    n=a
end
disp(['greater number is:',num2str(m)])
disp(['less number is:',num2str(n)])
c=n
    if mod(m,c)==0 && mod(n,c)==0
        break
    end
    c=c-1
end
disp(['B.M.M is:', num2str(c)])
```

CHAPTER 4

Simulink in MATLAB

4.1 What Is Simulink?

Simulink is a software package for modeling, simulating, and analyzing dynamical systems. It supports linear and nonlinear systems, modeled in continuous time, sampled time, or a hybrid of the two. Systems can also be multirate, i.e., have different parts that are sampled or updated at different rates[5].

For modeling, Simulink provides a graphical user interface (GUI) for building models as block diagrams, using click-and-drag mouse operations. With this interface, you can draw the models just as you would with pencil and paper (or as most textbooks depict them). This is a far cry from previous simulation packages that require you to formulate differential equations and difference equations in a language or program. Simulink includes a comprehensive block library of sinks, sources, linear and nonlinear components, and connectors. You can also customize and create your own blocks[11].

Models are hierarchical, so you can build models using both top-down and bottom-up approaches. You can view the system at a high level, then double-click on blocks to go down through the levels to see increasing levels of model detail. This approach provides insight into how amodel is organized and how its parts interact[5].

After you define a model, you can simulate it, using a choice of integration methods, either from the Simulink menus or by entering commands in MATLAB's command window. The menus are particularly convenient for interactive work, while the command-line approach is very useful for running a batch of simulations (for example, if you are doing Monte Carlo simulations or want to sweep a parameter across a range of values). Using scopes and other display blocks, you can see the simulation results while the simulation is running. In addition, you can change parameters and immediately see what happens, for "what if" exploration. The simulation results can be put in the MATLAB workspace for postprocessing and visualization[11].

Model analysis tools include linearization and trimming tools, which can be accessed from the MATLAB command line, plus the many tools in MATLAB and its application toolboxes. And because MATLAB and Simulink are integrated, you can simulate, analyze, and revise your models in either environment at any point.

4.2 Simulation

Simulink is an environment for simulation and model-based design for dynamic and embedded systems. It provides an interactive graphical environment and a customizable set of block libraries that let you design, simulate, implement, and test a variety of time-varying systems, including communications, controls, signal processing, video processing, and image processing[12].

Simulink offers:
- A quick way of develop your model in contrast to text based-programming language such as e.g., C.
- Simulink has integrated **solvers**. In text based-programming language such as e.g., C you need to write your own solver.

You start Simulink from the MATLAB:

Open MATLAB and select the Simulink icon in the Toolbar:

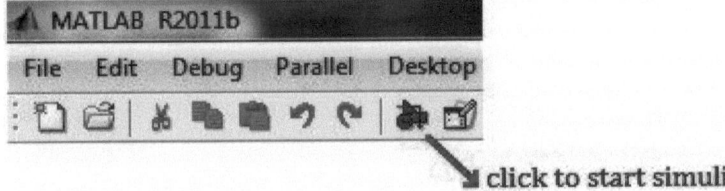

Fig. 4.1: Simulink icon

Or type "**Simulink**" in the Command window, like this:

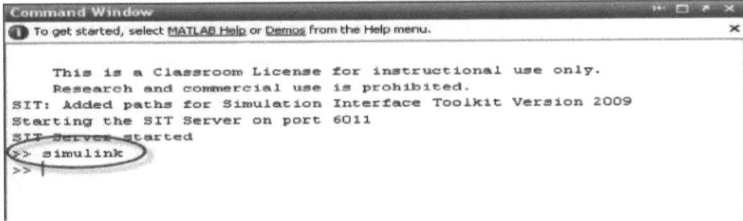

Fig. 4.2: Type Simulink in command window

Then the following window appears (**Simulink Library Browser**)[12]:

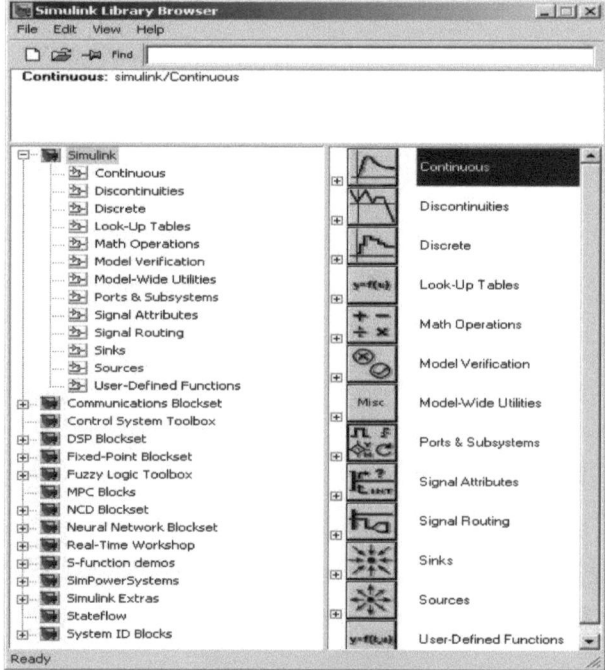

Fig. 4.3: Simulink library browser

Library Browser gives access to various standard or additional blocks that are used to build more complicated models

• ECEN2060 models will be constructed using standard Simulink blocks from the Simulink library

• Click File - New – Model (or Ctrl-N) to start a new model window

New Model button ⟶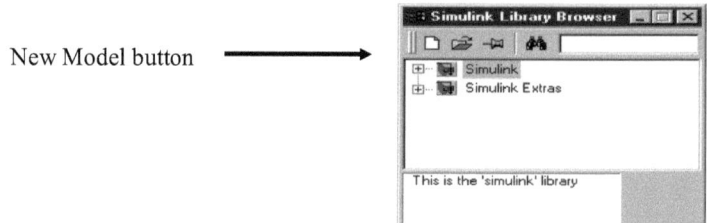

Simulink opens a new model window.

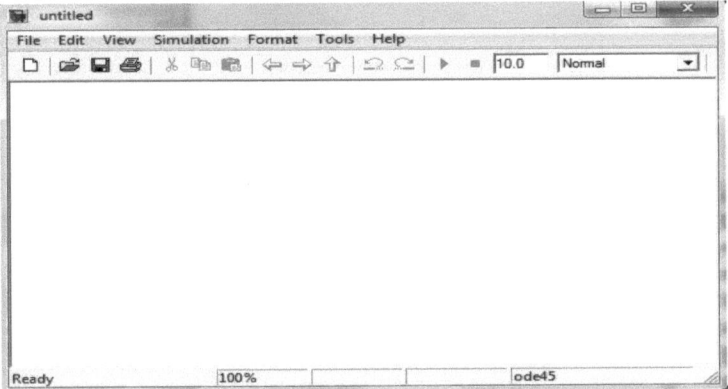

Fig. 4.4: Open new model for simulation

You might want to move the new model window to the right side of your screen so you can see its contents and the contents of block libraries at the same time. To create this model, you will need to copy blocks into the model from the following Simulink block libraries[12]:

•Sources library (the Sine Wave block)
•Sinks library (the Scope block)
•Continuous library (the Integrator block)
•Signals & Systems library (the Mux block)

You can copy a Sine Wave block from the Sources library, using the Library Browser (Windows only) or the Sources library window (UNIX or Windows)[12].

To copy the Sine Wave block from the Library Browser, first expand the library Browser tree to display the blocks in the Sources library. Do this by clicking first on the Simulink node to display the Sources node, then on the Sources node to display the Sources library blocks. Finally click on the Sine Wave node to select the Sine Wave block.Here is how the Library Browser should look after you have done this[12].

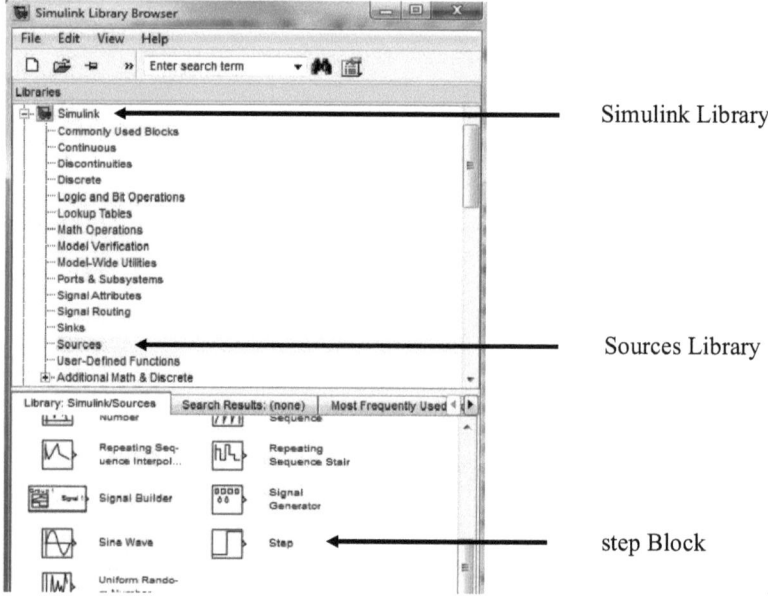

Fig. 4.5: Design new model[12]

Now drag the step Wave node from the browser and drop it in the model window. Simulink creates a copy of the SineWave block at the point where you dropped the node icon[12].

To copy the step block fromthe Sources library window, open the Sources window by double-clicking on the Sources icon in the Simulink library window. (On Windows, you can open the Simulink library window by right-clicking the Simulink node in the Library Browser and then clicking the resulting Open Library button.) Simulink displays the Sources library window[13].

Now drag the Sine Wave block from the Sources window to yourmodel window.

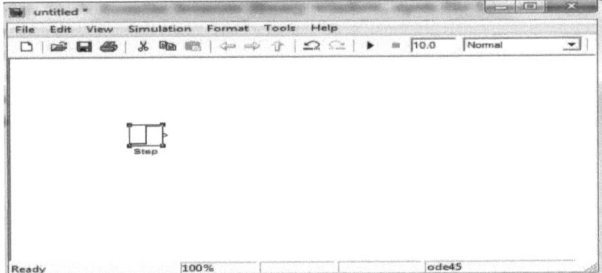

Fig. 4.6: Design Step source

For start, the simple model of is designed:

Example 1: Display the ingration of unit step function (step time =1):

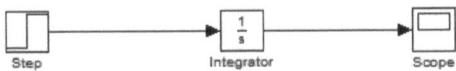

Fig. 4.7: Simulation of Unit step function

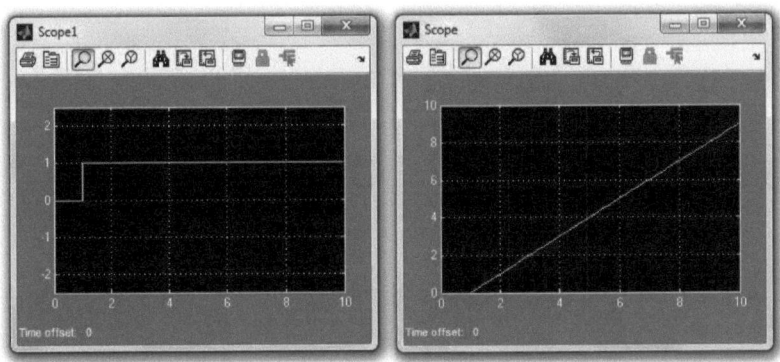

Fig. 4.8: Scope 1: Input of the Unit step function and Scope: The output of the Unit step function

- to select the mesearment tools, we must use the following path:

  library browser ⟶ simpower system ⟶ measurment

- To measure current and voltage, Current measurment and voltage measurment are common devices.
- In each electrical circuit model, Powergui must be exited that it can help to run the model.

Example 2: Measure voltage and cureent through a 10 Ω resistance that the voltage source is 25 V:

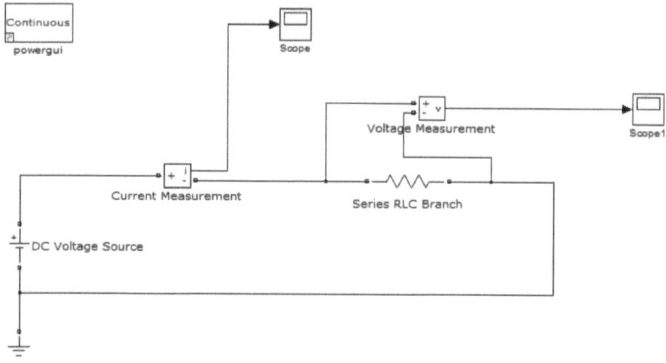

Fig. 4.9: Simulation of Example 2

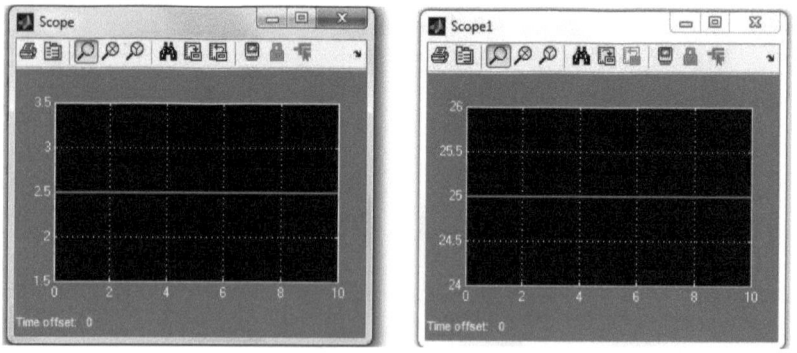

Fig. 4.10: Scope: Input of Example 2 and Scope 1 output of Example 2

After entering blocks to the model window, Settings of each block can be done by double-clicking on the desired block. For example, if we want to set one RLC branch on the desired value, by left double click on thos block and selecting R option, we can set the desired value such as the following shape:

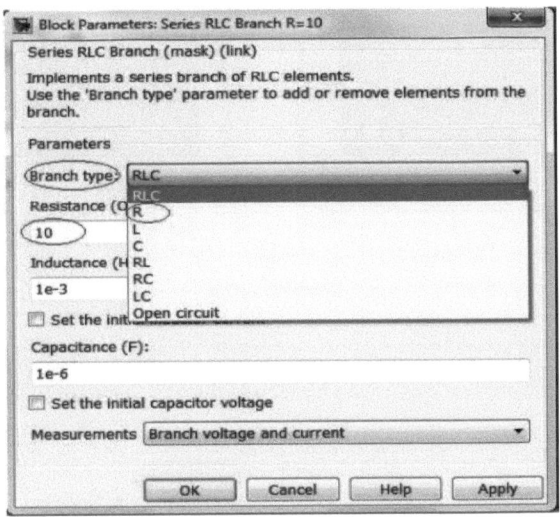

Fig. 4.11: Setting RLC branch on the desired value

Example 3: Obtain the capacitior charging curve (Vin=5v, C=0.1 µf , R=1Ω):

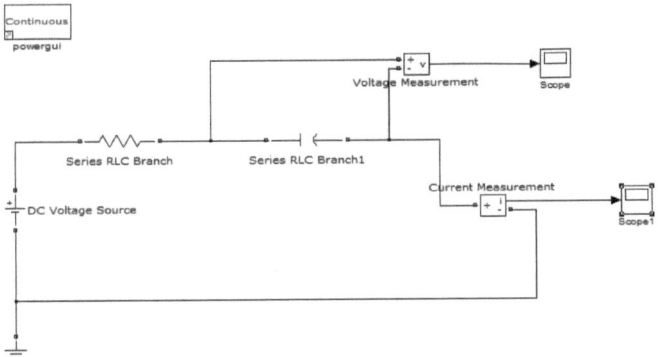

Fig. 4.12: Simulation of capacitior charging curve

55

Output waveforms from two scopes are shown as following shapes:

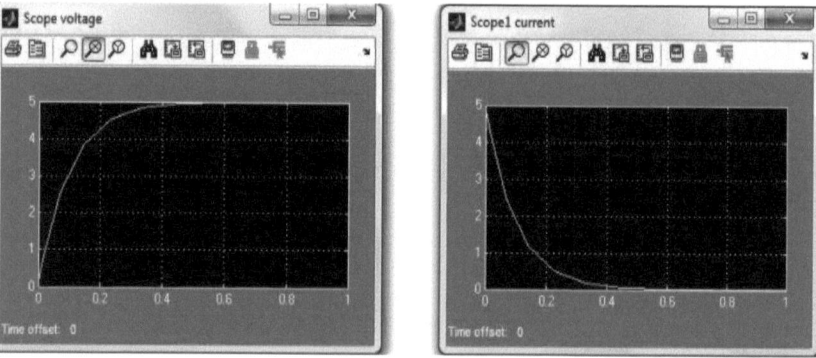

Fig. 4.13: Scope: Voltage of capacitior charging curve and Scope 1: Current of capacitor charging curve

We can use of one scope instead of two scopes, by changing the input of scope into two inputs. Therefore two curves can be seen in the one scope for this mean, we can use of follow method:

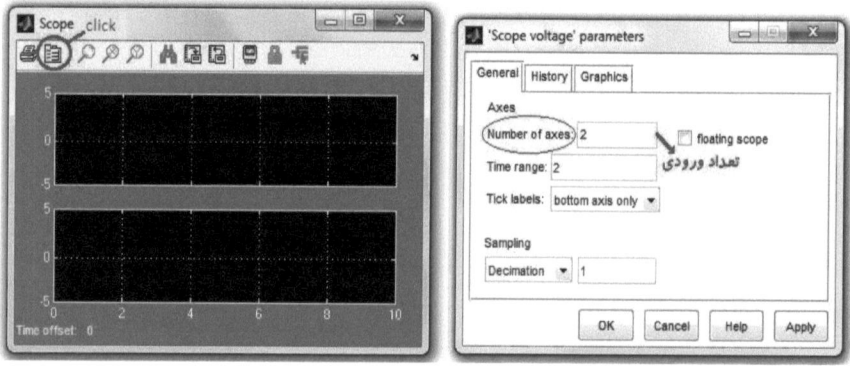

Fig. 4.14: Using one scope instead of two scopes

Also, because we want to start the charging curve from zero moment, we need to set the circuit that curve started from the zero moment. Therefore, initial value of capacitor must be set to zero value. The below shape shows this method.

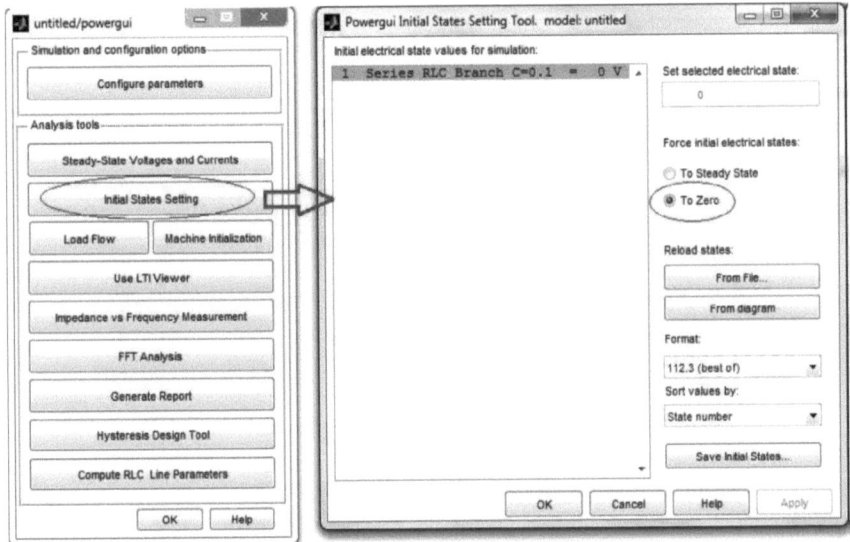

Fig. 4.15: Changing the value of capacitor

Example 4: Display The voltage and current computation for a three phase ohmic-inductive load, in the following figures:

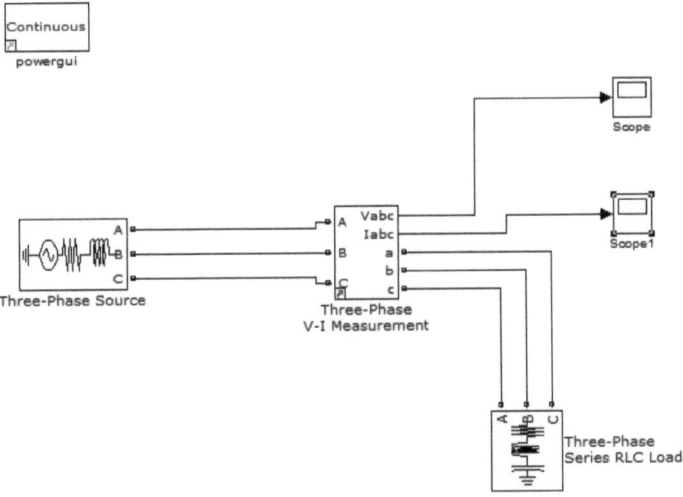

Fig. 4.16: Simulation of three phase ohmic-inductive load

Output waveforms from current and voltage measurments scopes are shown as following shapes:

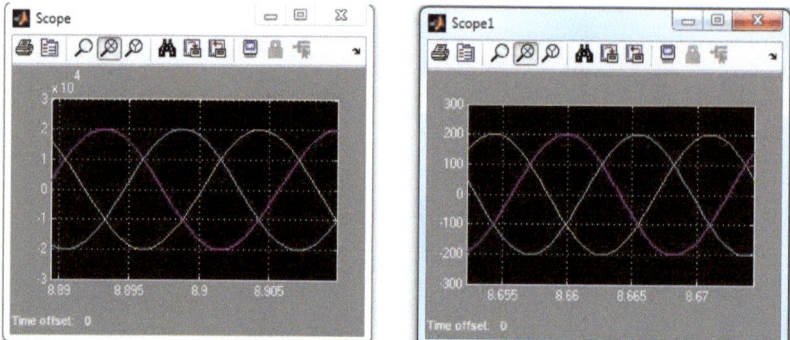

Fig. 4.17: Scope: The output voltage of three phase ohmic- inductive load and scope 1: The output current of three phase ohmic-inductive

Example 5: Design Three-phase bridge rectifier and display the input and output waveforms in the seperated shapes?

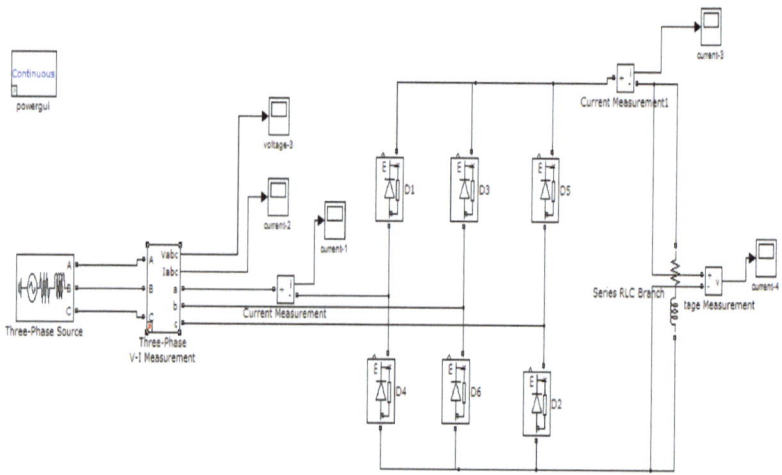

Fig. 4.18: Design of Three-phase bridge rectifier with MATLAB simulation

The input waveforms of voltage and current to the three-phase bridge rectifier:

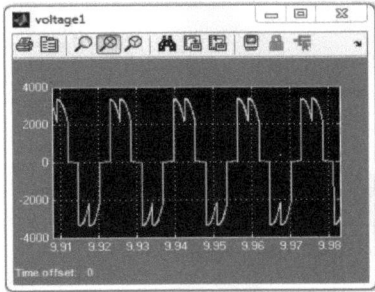 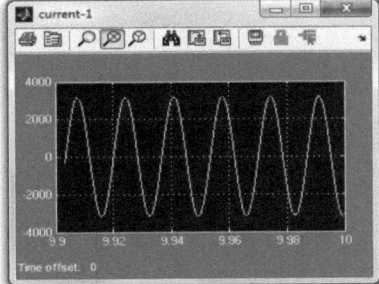

Fig. 4.19: The input waveforms of Voltage and current of the three-phase bridge rectifier

The output waveforms of voltage and current to the three-phase bridge rectifier:

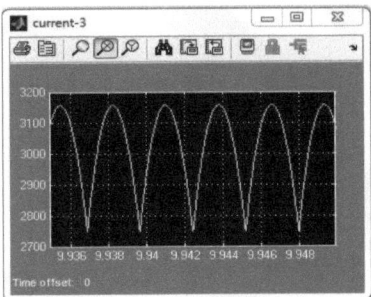 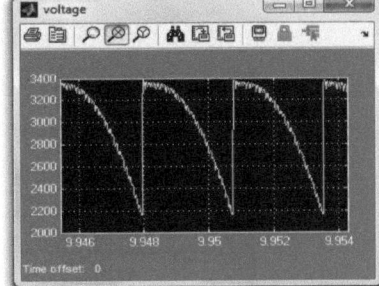

Fig. 4.20: The output waveforms of Voltage and current of the three-phase bridge rectifier

Example 6:

For the circuit of Figure 4.21, find the value of the current after combining the voltage sources to a single voltage source and the resistances to a single[13]:

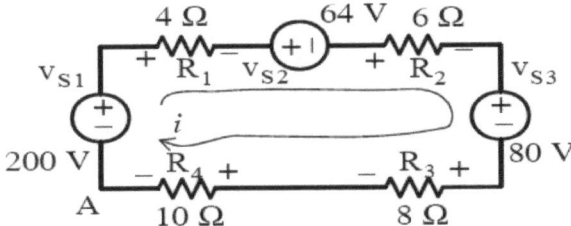

Fig. 4.21: 6

Solution:

we apply Ohm's law. Then,

$$i = \frac{\Sigma V}{\Sigma R} = \frac{200-(64+80)}{28} = \frac{56}{28} = 2\,A$$

Next, we consider the case where resistors are connected in parallel as shown in Figure 4.22:

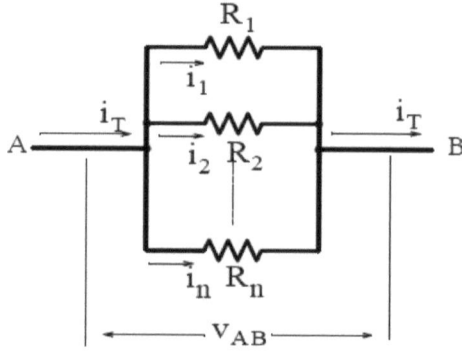

Fig. 4.22: Addition of resistance parallel

$i_T = i_1 + i_2 + \cdots + i_n$

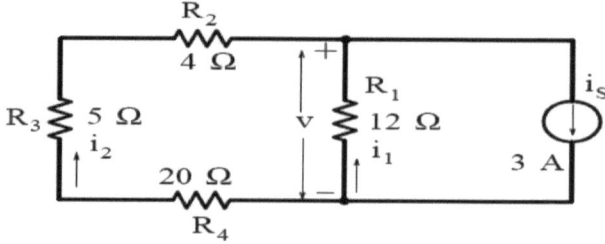

Fig. 4.23: Application of current division expressions for the circuit of Example 6

and observing the passive sign convention, the voltage is

$$v = -i_1 R_1 = \frac{-87}{41} \times 12 = -\frac{1044}{41}\,V$$

V=-25.46v

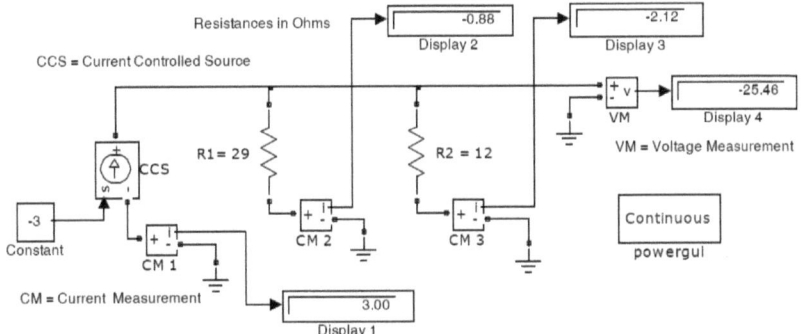

Fig. 4.24: Simulink / SimPower Systems model for Example 6

Example 7: Design the below circuit and display the V_1, V_2, V_3 in each nodes:

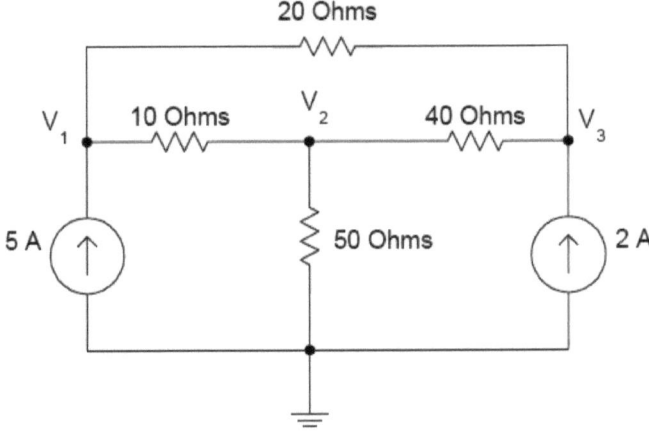

Fig. 4.25: Circuit for example 7

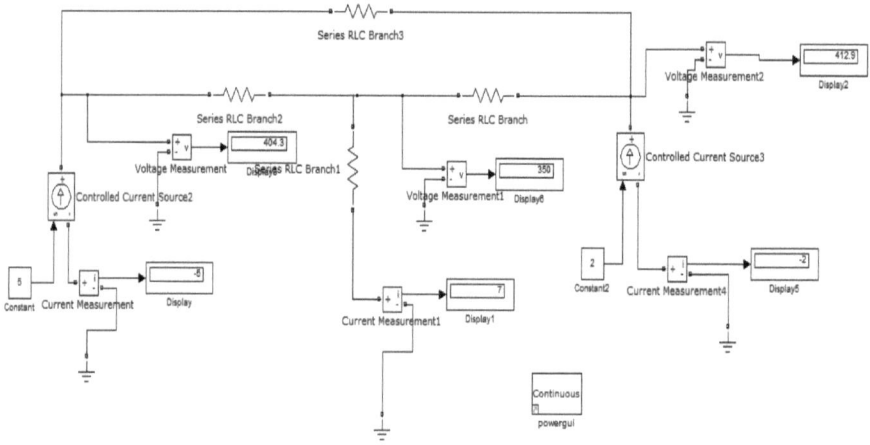

Fig. 4.26: Simulink / SimPower Systems model for Example 7

By simulation with MATLAB, the values of the V_1, V_2, V_3 **in each nodes**:

V=

404.3

350

412.9

CHAPTER 5

DC Circuit Analysis using MATLAB

5.1 DC Analysis

5.1.1 Node Analysis

''Kirchhoff's current law states that for any electrical circuit, the algebraic sum of all the currents at any node in the circuit equals zero. In nodal analysis, if there are n nodes in a circuit, and we select a reference node, the other nodes can be numbered from *V1* through *Vn-1*. With one node selected as the reference node, there will be n-1 independent equations. If we assume that the admittance between nodes i and j is given as Yij , we can write the nodal equations''[14]:

$$Y_{11}V_1 + Y_{12}V_2 + \cdots + Y_{1m}V_{1m} = \sum I_1$$
$$Y_{21}V_1 + Y_{22}V_2 + \cdots + Y_{2m}V_{2m} = \sum I_2$$
$$Y_{m1}V_1 + Y_{m2}V_2 + \cdots + Y_{mm}V_m = \sum I_m$$

(5.1)

Equation (5.1) can be expressed in matrix form as:

$$[Y][V]=[I] \tag{5.2}$$

The solution of the above equation is:

$$[V]=[Y]^{-1}[I] \tag{5.3}$$

Where

$[Y]^{-1}$ is an inverse $[Y]$

In MATLAB, we can compute [V] by using the command[14]

$$V=inv(Y)*I \tag{5.4}$$

Where

inv(Y) is the inverse of matrix Y

The matrix left and right divisions can also be used to obtain the nodal voltages. The following MATLAB commands can be used to find the matrix [V]

$V = I/Y$ (5.5)

or

$V = Y/I$ (5.6)

The solutions obtained from Equations (5.4) to (5.6) will be the same, provided the system is not ill-conditioned. The following two examples illustrate the use of MATLAB for solving nodal voltages of electrical circuits[14].

Example 5.1:

For the circuit shown below, find the nodal voltages V_1, V_2, and V_3[14].

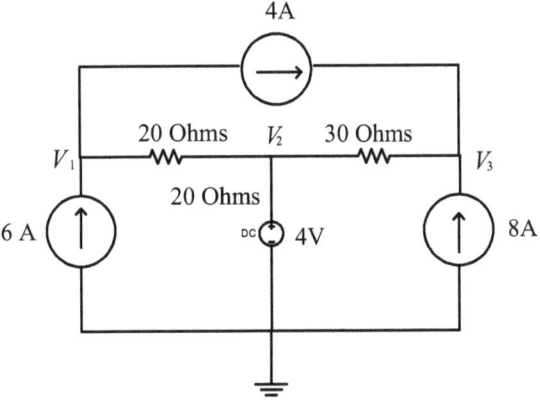

Fig. 5.1 Circuit with Nodal Voltages

Solution:

Using KCL and assuming that the currents leaving a node are positive, we have
For node 1,

$$\frac{V_1 - V_2}{20} - 6 + 4 = 0$$

$$V_1 - V_2 = 40$$

At node 2,

$$V_2 = 4$$

At node 3;
$$\frac{V_3 - V_2}{30} - 12 = 0$$
$$V_3 - V_2 - 360 = 0$$

In matrix form, we have

$$\begin{bmatrix} 1 & -1 & 0 \\ 0 & 1 & 0 \\ 0 & -1 & 1 \end{bmatrix} \begin{bmatrix} V_1 \\ V_2 \\ V_3 \end{bmatrix} = \begin{bmatrix} 40 \\ 4 \\ 360 \end{bmatrix}$$

The MATLAB program for solving the nodal voltages is:

Y = [1 -1 0;

0 1 0;

0 -1 1];

I = [40;

4;

360];

v = inv(Y)*I

The results obtained from MATLAB are

Nodal voltages V1, V2 and V3,

v =

44

4

364

5.1.2 Loop Analysis

''Loop analysis is a method for obtaining loop currents. The technique uses Kirchoff voltage law (KVL) to write a set of independent simultaneous equations. The Kirchoff voltage law states that the algebraic sum of all the voltages around any closed path in a circuit equals zero''[14].

In loop analysis, we want to obtain current from a set of simultaneous equations. The latter equations are easily set up if the circuit can be drawn in planar fashion. This implies that a set of simultaneous equations can be obtained if the circuit can be redrawn without crossovers[14].

For a planar circuit with n-meshes, the KVL can be used to write equations for each mesh that does not contain a dependent or independent current source. Using KVL and writing equations for each mesh, the resulting equations will have the general form[14]:

$$Z_{11}V_1 + Z_{12}V_2 + \cdots + Z_{1n}V_{1n} = \sum V_1$$
$$Z_{21}V_1 + Z_{22}V_2 + \cdots + Z_{2n}V_{2n} = \sum V_2$$
$$Z_{n1}V_1 + Z_{n2}V_2 + \cdots + Z_{nn}V_n = \sum V_n$$

(5.7)

Where

$I_1, I_2, I_3 \cdots I_n$ are the unknown currents for meshes 1 through n.

$Z_{11}, Z_{22}, Z_{33} \cdots Z_{nn}$ are the impedance for each mesh through which individual current flows[14].

$\sum V_n$ is the algebraic sum of the voltage sources in mesh n.

Equation (5.8) can be expressed in matrix form as

[Z][I]=[V] (5.8)

The solution to Equation (5.9) is

[I]=[V] [Z]$^{-1}$ (5.9)

In MATLAB, we can compute [I] by using the command

$I = inv(Z) * V$ (5.10)

Where

$inv(Z)$ is the inverse of the matrix Z The matrix left and right divisions can also be used to obtain the loop currents[14].

Thus, the current I can be obtained by the MATLAB commands

$$I = V/Z \qquad (5.11)$$

or

$$I = Z\backslash V \qquad (5.12)$$

As mentioned earlier, Equations (5.10) to (5.12) will give the same results, provided the circuit is not ill-conditioned. The following examples illustrate the use of MATLAB for loop analysis[14].

Example 5.2:

Use the mesh analysis to find the current flowing each branch. In addition, find the power supplied by the 8 voltage source[14].

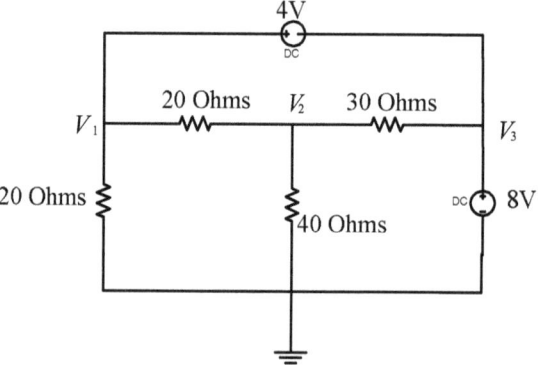

Fig. 5.2: Bridge Circuit

Solution:

Using loop analysis and designating the loop currents as $I_1, I_2, I_3, \ldots, I_n$, we obtain the following figure.

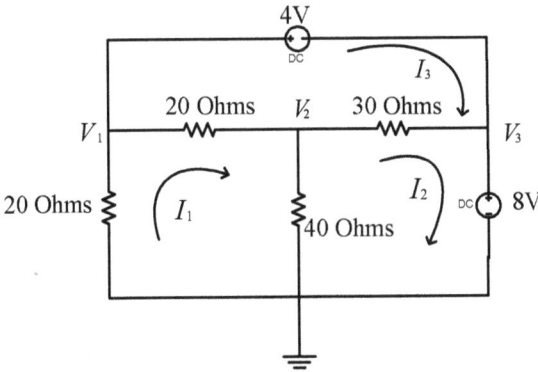

Fig. 5.3: Bridge Circuit with Loop Currents

Power supplied by the source is $P = 8I_2$.

The loop equations are

Loop 1,

$20I_1 + 20(I_1 - I_2) + 40(I_1 - I_3) = 0$
$80I_1 - 20I_2 - 40I_3 = 0$

Loop 2,

$4 + 30(I_2 - I_3) + 20(I_2 - I_1) = 0$
$-20I_1 + 50I_2 - 30I_3 = -4$

Loop 3,

$30(I_3 - I_2) + 8 + 40(I_3 - I_1) = 0$
$-40I_1 - 30I_2 + 70I_3 = -8$

In matrix form,

$$\begin{bmatrix} 80 & -20 & -40 \\ -20 & 50 & -30 \\ -40 & -30 & 70 \end{bmatrix} \begin{bmatrix} I_1 \\ I_2 \\ I_3 \end{bmatrix} = \begin{bmatrix} 0 \\ -4 \\ -8 \end{bmatrix}$$

The MATLAB program for solving the loop currents I_1, I_2, I_3, the current I and the power supplied by the 10-volt source is:

Z = [80 -20 -40;
-20 50 -30;
-40 -30 70];
V = [0; -4; -8]';
I = inv(Z)*V;
I =
   -0.600000000000000
   -0.800000000000000
PS = I(2)*8;

MATLAB answers for the power supplied by 10V source is -6.400000000 watts.

Example 5.4:

Nodal analysis provides a general procedure for analyzing circuits using node voltage as the circuit variables. The nodes of the circuit are the places where they are connected together. The circuit shown in figure 3.0 has three nodes. Node 3 is the reference node or the datum node.

Analyzing the connected circuit containing n nodes will require n-1 KCL equations. The KCL equations are obtained from each node with exception of the reference node or the datum node. Following are steps to determine the nodal voltages:

- Express element current as function of the nodal voltage
- Apply KCL to each node with the exception of reference node
- Solve the resulting simultaneous equations to obtain the unknown voltages.

The resulting simultaneous can be solved efficiently and accurately by using the circuit equation solver or we can use matrix method in MATLAB. The KCL equations can also be solved by using a calculator but this method is not as efficient as the MATLAB.

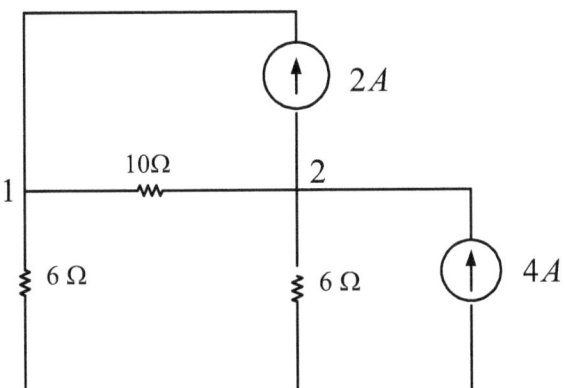

Fig. 5.4: The circuit of example 5.4

By applying KCL:

node 1:

$$\frac{V_1}{6} + \frac{V_1 - V_2}{10} - 2 = 0$$

$$0.167V_1 + 0.1V_1 - 0.1V_2 = 2$$

$$0.267V_1 - 0.1V_2 = 2$$

node 2:

$$\frac{V_2}{6} + \frac{V_2 - V_1}{10} - 4 + 2 = 0$$

$$0.167V_1 - 0.1V_1 + 0.1V_2 = 2$$

$$0.067V_1 + 0.1V_2 = 2$$

We can also use the matrix method to solve for v1 and v2.

$$\begin{bmatrix} 0.267 & -0.1 \\ 0.067 & 0.1 \end{bmatrix} \begin{bmatrix} V_1 \\ V_2 \end{bmatrix} = \begin{bmatrix} 2 \\ 2 \end{bmatrix}$$

The MATLAB program for solving the nodal voltages is:

Y = [0.267 -0.1;0.067 0.1];

I = [2;

2];

v = inv(Y)*I

The results obtained from MATLAB are Nodal voltages V1, V2 and V3,

v=

11.976047904191617

Simulink helps simulate the circuit to compute nodal voltages. We can model the circuit in figure 5.4 to determine the nodal voltages at node 1 and node 2. Following are the modelling steps:

- Model the resistors through the Simulink block (SimPowerSystems--Elements--Series RLC Branch. (Set the resistance to the given value, set L=0 and C=inf).
- Model the current source through the Simulink block (SimPowerSystems--Electrical Sources--Controlled Current Source).
- Model the constant for the current source through (Simulink--Sources--Constant).
- Model the voltage measurement through (SimPowerSystems--Measurements--Voltage Measurement).
- Block name display can be invoked through (Simulink--Sinks--Display).
- Nodes can be assigned by invoking the neutral through (SimPowerSystems--Elements---Neutral. The node number can be changed by double click on the neutral.
- Continuous power GUI must be embedded in the circuit. (SimPowerSystems—powerGUI.[15]

A Model of the circuit for the nodal analysis to compute voltages v1 and v2 is shown in figure 5.5.

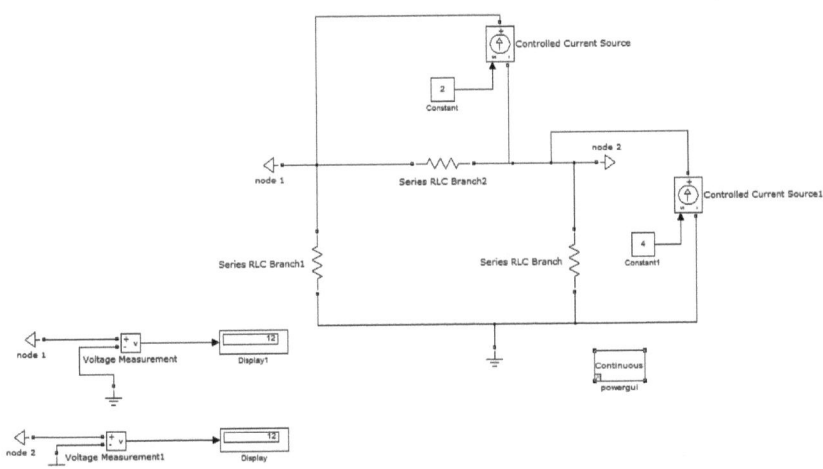

Fig. 5.5: Mode of the circuit of figure 5.4 for the node voltage analysis

References

[1] Gerritsen, M. (Sept, 2006). A brief introduction to MATLAB. Linear Algebra with Application Engineering Computations.

[2] Attaway, S. (June 2013). A Practical Introduction to Programming and Problem solving. Science direct.

[3] Houcque, D. (August 2005). Introduction To Matlab For Engineering Students. Northwestern University.

[4] Kalnins, L. (February 2010). MATLAB Basics.

[5] Brian R. Hunt, R. L. (2001). A guide to MATLAB. Published in the United States of America by Cambridge University Press, New York.

[6] Gockenbach, M. S. A Practical Introduction to Matlab. University of Michigan.

[7] Vick, B. MATLAB Commands and Functions. Mechanical Engineering Department Virginia Tech.

[8] Introduction to Plotting with Matlab. (1996). Math Sciences Computing Center University of Washington.

[9] Edward Neuman,"Programming with MATLAB"Department of Mathematics Southern Illinois University at Carbondale, Tutorial 2,1-46pp

[10] Edward Neuman"Using MATLAB in Linear Algebra" Department of Mathematics Southern Illinois University at Carbondale, 1-37 pp

[11] Simulink, Dynamic System Simulation for MATLAB. (1990-1999). The Math works Inc.

[12] Chee-Mun Ong," Dynamic Simulations of Electric Machinery: Using MATLAB/SIMULINK",1998, Prentice-Hall PTR press.

[13] John O. Attia, "Electronics And Circuit Analysis Using Matlab, 1999 By CRC Press LLC.

[14] Steven.T.Karris,"Circuit analysis I", 2009, Orchard Publications.

[15] A. Yousuf, M.A. Mustafa, W. Lehman," Electric Circuit Analysis in MATLAB and Simulink", 2014,121st ASEE Annual conference &Exposition, 1-12pp.

[16] Math Sciences Computing Center University of Washington" Introduction to Plotting with Matlab" September, 1996